BIG BLUE

***WRITTEN* BY**

ANNE E. REARDON

AF882559

PUBLISHED BY:

Amazon KDP

Ingramspark

USA

Other books written by Anne E. Reardon.

THE STORY Of AUTUMN - a children's story.

A NICE CUP Of TEA - an adult horror.

ABOUT THIS AUTHOR

Anne E. Reardon lives in Jensen Beach, Florida. Her hope is to engage people's curiosity about the ocean and share the importance of its necessity before it is too late

Please consider writing a review on Amazon.

Your input matters.

Thank you.

First Edition

ISBN Soft Copy 978-1-7320197-7-5

ISBN E-book. 978-1-7320197-8-2

FIRST EDITION

DEDICATION

Thanks to all the wonderful people in my community
who encouraged me to write about our ocean.!
Here's hoping it will last forever!

ACKNOWLEDGMENTS

Special thanks to:

All the people who are working so hard to save our oceans.

Sheila, whose input is always appreciated.

Kim, (my daughter) for her pure honesty.

To Hank and Shelly, my encouraging island friends.

To Pat who edits it all (thank heaven).

INTRODUCTION

First, let me thank you for taking the time to find out about what's going on and why I need your help. I will give you some history so you can better understand why this is an urgent matter! We don't have a lot of time to do what must be done if I'm going to survive, and in turn help you survive.

I cover seventy (70) percent of the Earth's surface and hold ninety-seven (97) percent of the Earth's water. I actually play a very important part in Earth's climate. I produce over half of the world's oxygen and I absorb fifty (50) times more carbon dioxide than Earth's atmosphere. I am not bragging! I want to make you aware of the facts and how different your existence will be if my destruction is not stopped. And, believe me, that is exactly what is happening! Without major change, my demise will be inevitable.

CONTENTS

CHAPTER 1

BIG BLUE

I wish you could have seen me in the beginning, I was *knock your socks off* awesome! Now I AM bragging! I had mountains that were covered with beautiful live coral and sponges, and healthy green algae that looked beautiful. The colors of the reefs were incredible, and the life teeming on them was, in and of itself, breathtaking. I was so happy to have so much to share with you.

I still have some great surprises that I've been saving for you, in what you call the *Mariana Trench* (it's about seven (7) miles deep). There are some species of ocean life you have never seen. I hope we can join forces and stop the destruction that is occurring daily, so some of the beauty and life will be left for your discovery.

There are seven hundred thousand (700,000) to one- million (1,000,000) different species of sea life that live within my oceans. Though you have seen much of it, you still have a lot more to discover! It honestly never occurred to me that it wouldn't be here for you forever, or that people would not care about the health and the preservation of nature. I believed we would always take good care of each other. My goal was to continue to have a beautiful place to share with you, and continue to provide the fresh air and water you require for survival. I watched with great respect as you progressed by using me to get you to different parts of the Earth that were now raised up above me. I was excited to know you were curious about me and wanted to know more. I was happy to help sustain you with food as you learned to catch fish, and to aid in your travels. There was a time when we had a mutual respect. Even when you let your waste drop into me, I figured out how to incorporate it into my growth pattern. You would throw your leftover food into me and the sea life actually enjoyed the diversity. Truth be known, they still do enjoy the food. Back then, you took only what you needed to survive and left me with a little cleaning, nothing I couldn't handle, so I could continue to provide for you. I enjoyed what was a great symbiotic relationship.

As you started diving into the water, even though you had nothing to assist you and you couldn't stay under too long, you seemed to love what you saw. And I loved seeing you! I liked the way you wanted to touch everything, and I could see by the look on your face you were captivated by the colorful reefs. I hoped you knew that the sponges and coral were alive. As time went by, you began to do what I was told you called snorkeling. It was great to have you be able to spend so much more time with me and get a bit more adventurous in going a little deeper. And little deeper you would go. Then it happened! I would never have recognized you had it not been for your two arms and two legs. You had your entire

body covered with stuff attached to you, like ropes and a helmet with a tube on your head. You were amazing! You spent so much time and descended so far, I was in awe of you!

That's when I decided it would be wonderful to let you see awesome things in my depths. Eventually, you just wore a mask over your eyes and nose and had a tube in your mouth attached to a tank or two on your back. Boy, did that give you some freedom. I never tired of watching you explore and seeing the excitement when you made a new discovery. I always held you close, and tried to make you feel safe.

I felt bad that you seemed to learn everything the hard way. In the beginning, as you went deeper and deeper, you didn't realize how much the pressure change, would be difficult for your body. That's because in comparison, to fish, you had so little body pressure above the water. What you call "the bends" was a painful learning experience, I'm sure. But you kept trying and learning, and like all things, you overcame those problems. That's what gives me hope that you can help with what's going on now.

Much of the sea life decided early on to become what you refer to as bottom dwellers. They saw that you began to take more and more of them from their habitat. Fishing became a sport; it was no longer to help maintain life. It was just for fun, and apparently you are a very competitive life form, as it

seemed the "bigger the better." As I understand it, when you hunt on land you actually capture the species that are not as healthy as they use to be, so in effect you promote *survival of the fittest.* It actually works the same way down here. The species that are the weakest, old, or unwell are the easiest catch. Most of the healthy ones avoid human contact and aren't easily deceived. I have to admit, you do get pretty clever in your bait and attraction methods. As always, you impress me with your ingenuity. That's the main reason I'm asking for your help now. I believe from the core of my being that you can solve this crisis before it is beyond reversal. I pray that your apparently massive needs for food and sport can be somehow quenched with other sources from on land.

Your military has left some sounding devices that send signals that help keep you safe, I guess. We have to listen to these darn *Pings* that make no sense to us at all. The residents here wish it was a different frequency, not so close to how they communicate.

It's hard for me to understand, since I reside over so much of the Earth and all the life I support shares the entire ocean, why you seem to have divided up the land and have small sections of it where different people live. Strange concept for me. But, compared to the other problems, that's nothing. I'm just mentioning the things that have been changed by your design… some good, some not so good.

I truly don't want this to be a totally negative and depressing report. I promise I will give you the facts, and please know I believe in you and your amazing ability to solve problems. This is a big one, but you have proven in all areas to be great problem solvers. I have seen firsthand what you can do when you put your mind to it. You are truly amazing!

Right now, I would like to introduce you to Leakey, my favorite Leatherback Turtle. She will take it from here. Between her personal experiences and all her friends that report anything new to her, she has either heard all about it or seen it in person.

Leakey would never brag, so let me give you some of her statistics. She is almost seven (7) feet long and weighs about one-thousand five-hundred (1,500) pounds. She has descended from a line of Leatherbacks that were here when the dinosaurs roamed the Earth. Everything that her family line has experienced over the centuries is a part of Leakey's knowledge. What she sees now, and the information brought to her by many friends, is what she will share with you. Thank you for taking your time learn about what's going on; and I look forward to us working together.

CHAPTER 2

LEAKEY

Well hello! I'm Leakey and I am honored to have been chosen by Blue to tell you more about what's happening down here. I'm sure he chose me because I am known for being so nosey. And, I never stop asking questions until I get an answer that I can confirm is factual. If you have any doubts about what I am about to share with you, please do the same. Confirmation is a good thing. I have so many sources, including my wonderful shark, whale and octopus friends who travel all over the world and are good enough to look me up when they get back to my area. Oh, and I almost forgot, there's Pell my pelican friend who rides on so many ocean liners and fishing boats. He sees first hand what's going on and always reports back to me so I can spread the word. Just so you know, a history of our sea-life existence has been passed down through my lineage for eons. We are taught early on about any changes that might affect our existence and yours. You see, we need each other if we are all going to survive. Recently there have been red flags raised in our existence. There is a serious contamination problem, and without the help of the people living on land all over the world, there will be no solution. You humans need to make some serious changes now. My hope is that some of the ideas you have might help stop this disaster that is in the making. For us who live in the sea, this is our

home; we have no way of eliminating the problem without you. For example, I have a turtle friend who had a plastic thing stuck around her neck. I found out it is called a *six-pack holder.* It took us almost all day to get it off, and she was left with an awful scar. There are other fish friends who search for food along reefs and end up eating plastic bags, or worse yet, plastic wrap, that is completely invisible in water. It's easily ingested along with their normal diet, and ends up in their stomachs. Or, because we can't see it, fish swim into it and what it does to them is awful! If it doesn't suffocate them, it shuts down their digestive tract; they are starved and poisoned all at the same time. Very sad to see.

There are also so many chemicals the fish ingest just by being exposed to them in the water. But more about that later.

Then of course there is me. Why is my name Leakey? Well, it's a bit embarrassing, but I think it's important for you to know how items that find their way into our ocean can change a life forever. In my case, it was what you call a straw. I had no idea it was sticking up in the coral where I liked to play. I was scratching my hind leg on some coral and I felt a small pinch; then suddenly I had a poo trail following me. I can't begin to tell you how embarrassing that was! Anyway, my friends worked hard to rid me of the straw. Unfortunately, they were not able to dislodge it. But, thanks to a barracuda friend with

very sharp teeth, we were able to snip it off close to the point of penetration and squeeze the end of it so not too much stuff is released… thus the name Leakey. So obviously I know firsthand what I'm talking about.

CHAPTER 3

WHAT YOU NEED TO KNOW

The first thing I want to do is to thank you for taking the time to find out what is going on. It really does matter, and without each and every one of you being informed, both we in the ocean, and you on the land are at risk.

Special thanks to all of you who are working to clean up the rivers, lakes and oceans. You have your work cut out for you, and you are showing yourself to be just wonderful for the task at hand. Congratulations on getting the world involved, this isn't a small local problem… it is everywhere throughout this Earth!

My hope is that when people are done reading this book they will contact others, or that some of you, will start your own awareness groups or recycling programs that encourage or promote less pollution. If everyone works together they can make so many good things happen! There isn't any one group of us that hasn't been compromised in one way or another by the stuff that gets put into our water.

More importantly, people need to know the ocean is the life-support system for the land portion of the planet. The ocean not only provides food to thousands of people throughout the world, it is also what gives the entire population fresh drinking water.

The ocean and the sun are the two forces that create the global circulation system. They bring

the water from the land to the sea, then up into the atmosphere, and back to the land again in the form of rain. Without the ocean you can say good-by to fresh water, and thus say good-by to life. But, it isn't just the ocean's water that takes care of this. It's also the plant life, namely the plankton in the ocean, which provides so much of what you need to survive.

Another big problem which doesn't seem to be talked about, is the animal-killing industry. It is robbing the sea's oxygen production for short-term profits. Let me explain. Whales and, fish are absolutely necessary to fertilize plant life in the ocean. The whales are the primary species that are able to fertilize *phytoplankton*, the tiny little creatures that live in the the ocean and whose job is to remove carbon dioxide and provide oxygen. The carbon becomes part of their bodies. This is the major contribution of plankton, along with forming the basis for the entire ocean food chain.

Okay, what I'm about to explain to you now is fundamental and of the greatest importance. When I asked my pelican friend Pell what was going on, he told me about the massive amounts of whales being caught and killed. I knew you could not possibly know how your existence is dependent on whales.

One Blue Whale defecates three (3) tons of nitrogen and iron-rich feces a day. That's six-thousand (6000) pounds of poop! I bet you're glad

they don't live on land! However, this is what provides the nutrients to the phytoplankton that in turn feed the zooplankton, the fish, and ultimately everything that lives in the sea. In order to restore the diminished *phytoplankton* populations (the micro-organisms that help create your air) are disappearing from the reefs, we must restore the population of whales. We need to abolish the exploitation of bio-diversity in the ocean. Governments need to stop subsidizing commercial fishing operations and completely stop killing whales. Will this cost profits and jobs? YES! Unfortunately, the greed of some people have brought us to this dire situation. Ask yourself, ARE JOBS AND PROFITS really worth destroying the planet's LIFE-SUPPORT SYSTEM? I vote NO! People won't need jobs or money if they don't have air to breathe.

CHAPTER 4

WHAT CAN BE DONE

Think about this! More than five-hundred (500) people have traveled to outer space. Outer space has little to do with your day-to-day survival, right? In comparison, only a very small group have been to the deepest part of the ocean. This is the place that provides your daily oxygen and water, the place that is basic to your survival. People know more about the moon than they do the ocean floor. What you may not know is that the ocean survives just fine, in fact better, without people. Without the ocean however, people cannot exist.

Humans dump eight (8) million tons of plastic into the ocean each year. It's killing the fish that help keep you all alive. It is also going to inhibit the production of CO2, also what keeps you alive! This isn't a possibility, this is a fact!

There is a lot that can be done; however, time is running out! The world is literally at the 'fix it now or it won't matter' stage. Can we do it? YES WE CAN!

There are several good solutions. Do some searching, find what works for you, and become a part of the solution. You can start with doing simple things like not using plastic wrap. It is invisible in water and that's where it ends up. You can get wrap made of bee's wax. It's great stuff, all organic and reusable. Try buying your water by the gallon; you'll have to dispose of much less plastic. You and

each person in your family can each use one insulated container to drink out of, so you can eliminate the need for plastic bottles. Or, get a refrigerator that has filtered water all the time; no more plastic bottles for you.

Ocean pollution does effect all humans in one way or another, so please keep reading. I know statistics are boring, but it's important that you realize this is happening now. There is limited time to solve these problems before it is beyond rectification.

Most humans go through life having no idea how important the oceans are to their survival. I believe you will be better for knowing this information. So here are the facts. Seventy percent (70%) of the oxygen you breathe is produced by marine plants. Ninety-seven percent (97%) of the water supplied to the Earth is contained in the ocean. As much as thirty percent (30%) of the CO2 emissions that you produce are absorbed by the ocean. THIS IS THE IMPORTANT PART… the oceans are on the edge of a no-return disaster due to pollution by people! Millions of tons of debris is being dumped in the ocean constantly. This consists of metal cans and barrels full of toxic waste and plastic. In 2007, 2.12 billion tons of waste was dumped into the oceans. In 2010, two-hundred fifteen (215) metric tons of plastics were disposed of in the seas of the world.

It was discovered in 2014 that five-thousand (5,000) square miles of the Gulf of Mexico was devoid of life due to pollution. There are four-hundred five (405) of these same types of 'dead zones' that exist worldwide. Many are caused from nitrates that leach into groundwater supplies through agricultural runoff and find their way into the oceans. Nitrates exhaust oxygen in the water, and without oxygen ocean life dies. Relaxing the laws that govern off-shore drilling need to be stopped NOW! There is no more time; it cannot be put off another day. Tomorrow is NOW! Oceans are the one basic necessity to the survival of life on our planet that cannot be replaced with anything else. It is the foundation of Earth's eco system. If mankind is going to survive, you must have a compatible relationship with the oceans.

Now you may better understand why Big Blue couldn't tell you the story. He really doesn't want you to think he's a braggart. What he does now, and has always done, and will continue to do, is to work for what is best for our Earth and those who live on it. Unfortunately, all the space he occupies on this Earth seems to have simply provided more space for the dumping of waste. It seems business and money are the deciding factors, not the health of our existence. The ocean is being used as a personal dump for the world.

Please be advised, there are those among you who

honestly do not care about things beyond their own lives and immediate income. They will try to convince you that you need not be concerned, that these facts are not true. That this is a blatant lie. They seem to be compromising the future of mankind to make or save money now. Please think about it; what good will all that money be if there is no air to breathe, and no way to maintain life on Earth? This is not a 'maybe!' It's happening right in front of the eyes of the whole world!

Right now you're probably telling yourself that this has nothing to do with you. You don't litter and you don't pour toxins down your drain. So let me ask you, do you use plastic, Styrofoam, or aluminum? If so, how do you dispose of it? Let's do some comparisons. Cardboard takes about two (2) weeks to degrade in water and newspaper takes about six (6) weeks. Plastic takes four-hundred (400) years, foam takes fifty (50) years, Styrofoam takes eighty (80) years, and aluminum takes two-hundred (200) years. It all boils down to this… what you do on land determines the fate of what goes on offshore. Errors in waste disposal can have major impacts to human health as well as that of sea life.

The sea life takes in food or water that contains animal waste, and harmful chemicals such as pesticides, petroleum, or fertilizers. This happens because there is run-off due to watering, or heavy rainfall on farm land that goes into underground

streams, and eventually ends up in the ocean. Lead and other heavy metals are found in the polluted water and in the food chain of marine life. Marine life absorb these toxins which become part of their substance, then you buy and eat the fish. These toxins can and do cause hormonal problems, nervous system damage and kidney damage just to name a few. These chemicals that are so arbitrarily disposed of, or are allowed to enter the ocean, end up on YOUR plate! Chemicals are also blamed for diseases such as Parkinson's, Alzheimer's and heart disease.

Pollution on the beach can cause illness through physical contact or ingestion of water. I know, it's as though going to the beach is a hazard to your health; scary isn't it? So what to do? The reality is that cruise ships dump waste, and countries like China, Indonesia, The Philippines, Thailand and Vietnam are dumping more plastic into the ocean than the rest of the world combined. Maybe I'm an eternal optimist, but I really believe that if people around the globe are exposed to the truth of what's going on it *will* change. We in the ocean need you out there to broadcast the problem world-wide and to make changes in your own life.

I know you don't pollute on purpose, but do you have any idea what happens to the stuff you throw away? Pell, my pelican friend who I mentioned earlier, calls me up to the surface at least once a month to tell me about the situations he sees. Like

waste dumps that are burned and soaked so all the residue sinks into the land, making is way to underground waterways. Along with other stuff, Pell says this is causing us ocean dwellers real problems. We talk about all the chemicals filtering down through the Earth to the streams, lakes, rivers and finally ending up in the ocean. And they do a ton of damage along the way. Pell says it amazes him that most of waste dumps are near water, and wonders why this is so.

According to Pell. The stuff that's dumped from the ships way out at sea is shocking. No one's around to oversee; no one owns the ocean, so anything goes. I have racked my brain and I don't know what we can do about that. Pell says he has heard there are laws forbidding all the dumping, but apparently people aren't aware or maybe don't realize the danger. My hope is that by educating people they will on their own realize the damage isn't just to everyone else. It's affecting you, every single person living on the face of this Earth! No joke, there are NO EXCEPTIONS. No one will survive the death of the ocean. It is imperative that we broadcast this problem in every language throughout the Earth. You must all individually make changes about how to handle everyday items made of plastic, and how to dispose of cleaning products and medical hazardous waste. People worldwide must work together to enact and enforce laws governing this pollution.

WHAT CAN BE DONE

In 2019, plastic is still a problem, but the most toxic waste being dumped in the ocean is dredged material, industrial waste, radioactive waste and sewage sludge. In spite of the MPRSA (Marine Protection Research and Sanctuaries Act) passed in 1972, the contamination continues. This has to stop and it has to stop NOW! Literally, your life and mine depends on it, as does that of our entire ocean!

Okay, so we know there is a problem and we know it's worldwide. That makes it seem like an impossible task as far as finding a solution. IT ISN'T!!! YOU CAN DO THIS! There are groups right now that are working and are succeeding in their endeavors. There are already sixty (60) countries trying to correct processes that cause problems, and are working to remove plastic from the ocean.

CHAPTER 5

GO FOR IT

We just need to get the message to everyday people throughout this great world of ours.

The whole world must start to put things in place to stop dumping into the ocean. That's a good beginning. Maybe there could be trucks and manpower on the docks ready to unload debris from the ocean liners when they stop at ports along the way for the 'day trips.' On a positive note, this would create jobs! Or, why not give incentives to companies to dispose of or eliminate the waste in the healthiest possible way. Perhaps companies can change the chemicals used on farmlands. That way, when it rains, only good stuff ends up in the runoff, not toxins. People can stop putting harmful chemicals on plants and fields. Then only 'healthy' will be washed into the waters that at some point enter the ocean. Not to mention your food might be healthier.

Healthy disposal will help. Companies who recycle can start paying people who return plastics by the pound.

People can and will make a difference. Educating everyone about the definite destruction that is now creating full-blown health issues with long-term effects and even death will only help. They can stop global warming, yet another part of the problem. By stopping the dumping of plastic and other toxins into the ocean, they can clean up what is in there now.

How about using alternative energy sources instead of drilling for oil and natural gas. You have all the tools to make our Earth, and our ocean, once again symbiotic.

Talk about what you have learned; learn more and share your knowledge. That will help! Go online and type in '*names of organizations saving the ocean.*' Pick one and check it out. Become a volunteer, or start something of your own. Whatever you do will be a benefit to a problem that needs to be solved.

GO FOR IT!

Don't let the money make the decisions. Let's help the ocean be the magnificent supplier of life it was meant to be.

There are two-hundred six (206) countries in the world and only sixty (60) of them are involved in ocean clean up. Let's see if we can't increase that by one-hundred forty-six (146) more!

Blue, Pell, my many unnamed friends and I sincerely appreciate you reading about this problem and becoming a part of the solution.

Now you know what can be done. I hope together we can educate the world and make it happen.

May this not be THE END of this story, but rather a GREAT BEGINNING that ends, with a *happily ever after.*

NOTES

YOUR IDEAS

YOUR PLAN

Made in the USA
Monee, IL
06 November 2026

67950722R00018